WE CAN, WE WILL

READY AND FORWARD

BY

TROOPER HAROLD S. COLE

As Told To
Colonel (Retired) Dr. Gerald D. Curry,
U.S. Air Force

We Can, We Will, Ready and Forward

ISBN 978-0-9847742-7-2   Hardbound
ISBN 978-0-997353181   Softbound

Copyright © 2016 Trooper Harold S. Cole

Request for information should be addressed to:
Curry Brothers Marketing and Publishing Group
P.O. Box 247
Haymarket, VA 20168

All rights reserved. No part of this publication may be reproduced, stored in a retrieval system, or transmitted in any form or by any means, electronic, mechanical, photocopy, recording, or any other, except for brief quotations in printed reviews, without the prior permission of the publisher.

Cover designed by Aubria Hull

WE CAN, WE WILL

READY AND FORWARD

Buffalo Soldiers protecting the stagecoaches and taming the West

"Pride over Prejudice" by Rick Reeves

Dedication

This book is dedicated to men and women who serve in our nation's military, and continue to give so freely of themselves, especially the members of the Ninth and Tenth (Horse) Cavalry Association, my loving family, and beautiful wife Edwina. This book would not have been possible if it were not for the love and support shown throughout the years. Truly, each of you have sustained me by blessing my spirit and I will forever be indebted to you.

Acknowledgment

I would like to thank all the folks throughout my life who have played a positive role in supporting my career and life during the good and challenging times. I especially would offer a more formal appreciation to my father, George Leon Cole, who taught me the importance of hard work, caring for our family, and being there during the tough times. A very special thank you to my Dr. Redmond S. Oden, who was the Pastor at St. Catherine AME Zion Church, in New Rochelle, New York from 1936 to 1946. Pastor Dr. Oden taught me the value of spiritual grounding, which served as an anchor throughout my military career and life.

I would also like to recognize several people who played a pivotal role in shaping my life;

- My oldest sister Evelyn Cole
- My oldest brother James Cole
- Trooper Henri A. Legendre, 9th Cavalry, USA 1943
- First Sergeant Britton of Troop F, 9th Cavalry, USA 1942 – 1946
- Staff Sergeant Walls of Troop F, 9th Cavalry, USA 1942 – 1946
- Colonel (Retired) Franklin J. Henderson, 101st Airborne, 9th & 10th Horse Cavalry, USA
- Major General Gravette, First Black Commander for California National Guard
- Major General Fred A. Gordon, 61st Commandant of Cadets at West Point
- Colonel (Retired) Dr. Gerald D. Curry, USAF 1983 – 2010
- William H. Leckie, Author of Buffalo Soldiers
- Commander (Retired) Carlton Philpot Jr., USN
- My loving wife Edwina Cole

SPECIAL TRIBUTE

The good Lord has blessed me with a very special life-partner when he sent me Edwina Roberts-Cole as my wife. She is a very religious woman, takes great care of our family, active in community activities, loves gardening, and most importantly takes care of me. I will forever be indebted to her for all the love and support she has shown me over the years. Truly, she is heaven sent! God created an angle when he made Edwina. I will love you forever!

TABLE OF CONTENTS

Dedication	iv
Acknowledgment	v
Special Tribute	vi
Chapter 1. Early Days	1
Chapter 2. Standing on Shoulders	11
Chapter 3. Service Time	21
Chapter 4. Making History	27
Chapter 5. Family's Impact	35
Chapter 6. Fort Clark	43
Chapter 7. Fort Sill	55
Chapter 8. The 24th U.S. Infantry	65
Chapter 9. The 25th U.S. Infantry	71
Chapter 10. Remount	77
Medal of Honor Recipients	83
Fiddler's Green	89
Author's Biography	91

CHAPTER 1. THE EARLY DAYS

Trooper Harold S. Cole, Bracketville, TX at Fort Clark - 1943

I was born on October 28, 1924, in North Pelham, New York. My parents are George Leon Cole, and Arline Batman-Cole. We lived at 582 Seventh Avenue, in a three story house, which was considered a "cold water house," which basically means we had no hot water. There lived a family on each floor. We lived on the first floor. Our heat was a potbelly stove in the living room and a cooking wood stove in the kitchen. There was one room for Mom and Dad, one for the girls and one for the boys, one living room, dinning room, and a bath. An open porch in the front and a closed-in porch on the back, a full basement. A front yard and back yard. We lived right in back of the North Pelham Fire House. And had to deal with the annoying fire blast going off. What a noise! Besides fires, it went off every day at 8am, 12 noon, and again at 6pm. We had electric power from the time I was born, but you could see the gas fixture on the walls.

George Leon Cole (Sept. 17, 1891 - Nov. 10, 1977) & Arline Bateman Ford Cole (Nov. 29, 1893 - Jan. 16, 1967)

My father and mother both worked. They told me my father was well off before I was born, but that was past-tense for me. We were poor, from the time I can remember. The boys had to bring in a barrel of wood a day, all year and stack it in the cellar. There was a lot of open land we called "lots" and people could throw their ashes in the locations. We would go out early in the morning and shift the ashes for the coal that did not burn. We would get about a bushel full a day, that was good but it was cold and we would be wet to the skin. We had orchards, there were apples, grapes, peaches, cherries, plums, and berries. We had lakes and streams, parks pools, and play grounds. This is where a young child is rich. On my side of town there were only two Black families. The Cole's and the Millers, the rest were mostly Irish and Italians. Everybody knew everything about everyone. Later there was other black families, that moved in; the Joanstones, and Farrs. I did not know of any discrimination. My family did not teach it. We could go anywhere and any place the white people went. To include riding the bus, attending schools, theaters, places to eat, stores, and etc. We attended a black church in New Rochelle, and Mt. Vernon, New York.

Left -Evelyn Elizabeth Cole-Harris (May 13, 1914 - Feb. 24, 2004):
Husbands Wilford Lewen & Johnny Harris

Right-George Leon Cole Jr. (1915): Departed life as an infant

Evelyn, my oldest sister took care of us when my mother was working in those days. Whatever the oldest said was the law. But we were a good family. My father worked hard and played hard. He drank too much, but he took care of his family. Before my time, my brothers and sisters, said that he would do many wonderful things for them. To me, he was a very good father. I loved him. My mother was everyone's love. She was God-sent! Don't miss understand me, we had family fights and whatever families did to resolve issues. But we loved one another. We supported one another outside the house.

We had a railroad, that ran over our street, and a station was a block away. Up the street the fire chief, and his wife lived in a big white house, that had hundreds of cats all over the place. When a cat had a litter, she would get us to go under the house and get the litter before the mother would eat them.

The school, we attended was a block away, which was Hutchinson School. I had a hard time learning how to read and write. I think,

Left-James Bateman Cole (Oct 12, 1916 - Jun. 16, 1990): Wives - Ella Coleman, Marjorie Johnson, Harritt Cole, Estelle Johnson

Right-Cornealia Winifred Cole-Dance (Aug. 23, 1918 - Nov. 5, 2003): Husbands Kenneth Long, Robert Coleman, & Joseph Dance

I now understand why. There was a lack of teaching in our home. No one's fault, if you are trying to make a living for thirteen people. Also, a limit to their education, the only thing I wanted to do was get a job and help, my family. But I did pass my grades. I loved art and physical education. I was good, in math, but English was a hard subject for me. I got A's and D's, but I really think I needed motivation. I played basketball and baseball, on the school team.

Left- Doris Gwendolyn Cole (May 27, 1921 - Nov. 17, 1987): Husband - Walter Harris

Right-Andrew Watrus Cole (Feb. 22, 1923 - Nov. 5, 1997): Wife - Dorothy Britton

Early in the morning I would deliver milk, to individual class rooms. That was a good job. I joined the cub scouts, in the rich side of town. My mother got my uniforms from the people, she worked for. But it was hard to get the rest of the camping equipment and I lost out on a lot of the trips, then the pack moved to another area. That was my cub scout career. I wanted to be a Cub Scout, before I could never get to 4ft tall.

Left-Harold Sandford Cole (Oct. 28, 1924): Wives - Janet Gonsavles, Roberta Coleman, Vernell Bradly, & Edwina Roberts Cole

Right-Edith Leona Cole-Robinson (Feb. 8, 1927 - May 18, 2008): Husband - Peter Robinson

My older brother Jim, used to work at the Pelham Riding Academy, which was about a mile from our home. We used to go there and ride the horses to the watering trough. In the morning, noon, and night, I loved to ride those horses. My Grandmother, use to come visit us, and she was a wonderful lady. Whenever she came, all rules were over! Whatever us children wanted to do, we did! Also she would buy tons of food and good things. I remember her buying chickens and taking them in the back yard and ringing their necks. We called her "Mother Batman". My Grandfather was just the opposite. He would make you wash in cold water and eat hard bread, he was a very big and tall man.

Left-Donald Gilbert Cole (Apr 8, 1929 - Apr 8, 2013): Wives - Altorea Dallas, Irma Brown, Elizabeth Morgan, Mildred Barnes, Betty Die, Linda Beaty-Cole

Right-Keith Winfred Cole (Nov. 11, 1930 - Apr. 27, 1999): Wives - Dorothy Bedford, Jacquetta Peters-Cole

The school taught you how to be thrifty, we had a school bank account, but your parents controlled it. Also if you gave Mom some money to save for you, that was okay. But when you wanted it back, she would say, *"Did you eat last night, well that was your money"*. One day the first of January, I asked Mom for a nickel, she said, *"Harold I'm going to give you this nickel, but you don't ask me for another penny the rest of the year"*. My father was more generous, that is when he was outside the house. My brother, Andrew was a very honest person. There was nothing he would not do for you. We would get him in fights and do our work, send him on wild goose chases. But he would tell everything. He would help people work for no pay, he was loved by everyone. My brother Jim, (James Bateman Cole) was ten years older than me.

Barbara Jean Cole (1932) - Departed life as an infant

He was a giving and very likable person. All over Westchester, everyone knew Jim Cole, he was the best motorcycle rider I ever saw. He really liked me. He would leave me a five cent bag of candy at the store at least once a week. He bought me my first bicycle, and car. He was hard on clothes. He'd put on a new suit, and that next minute he would be under a car repairing it. Him and the suit would be greased and dirty. These two brothers were very instrumental in my childhood.

My father really was the protector of his family. No one could do anything bad to us. When I was ten, we moved from 4th Avenue, then to Fifth Avenue, in back of a bakery store front. There was one room in the Bakery and an apartment in the back, for us it was very nice, the wood and coal days were finally over.

Mr. George Leon Cole

George & Arline Cole

Mrs. Arline Bateman Cole

We moved to New Rochelle, when I was twelve years old. I was in junior high school. My sister Edith, caught up with me when we moved. She had been skipped two grades ahead. Pelham had better schools. I got a job working for the postal telegraph. I was the first black to work on that job. There was a lot of things before that, the guys had three black teams, clubs etc., they had the Oriole, and that the club I was in. The Black Yankees, the Westside and Hollow. These also, were different parts of New Rochelle. We held dances, plays, baseball, basketball, and had regular club meetings. I've had a good life, and I thank God every day!

SSGT Robert Powell

SSGT Robert Powell was awarded the Silver Star with the Ninth Infantry Division in Germany (deployed). SSGT Robert Powell, Company A, 47th Infantry Regiment, who distinguished himself by gallantry in action against the enemy on April 15, 1945. Near Hergenrath, Germany, the assault elements of the Infantry were subjected to intense enemy artillery machine gun and small arms fire. Of SSGT Powell's actions, the official Silver Star reads:

"Realizing the need for a rapid advance, Sgt. Powell immediately exposed himself to direct enemy observation and fire to single-handedly assault the enemy strong point. Although his carbine was shot from his hands, he unhestiantly obtain a hand grenade and lobbed into the position. Sgt. Powell's bold and fearless actions resulted in the neutralization of the enemy, contributing materially to the success of his operations."

SSGT Powell has been a member of the Ninth Division since March 12, 1945. At the time the Division was securing Manage Bridge heading across the Rhine River. He entered the service October 20, 1941 and trained at Fort Huachuca, Arizona. Previous to joining his present division, he was assigned to the 93rd Division, 3191st Quartermaster Service Company.

CHAPTER 2.
STANDING ON SHOULDERS

Morgan State Alumni (L-R) Gen Larry Ellis, Richard Robinson, Thomas Prather, Gen Kip Ward, Lt Gen Arthur J. Gregg

History of the Buffalo Soldier Story

Nearly sixteen months after the end of the Civil War, Section 3 of an Act of Congress entitled "An Act to increase and fix the Military Peace Establishment of the United States" authorized the formation of two regiments of cavalry composed of "colored" men. The act was approved on July 28, 1866. On September 21, 1866, the 9th Cavalry Regiment was activated at Greenville, Louisiana, and the 10th Cavalry Regiment was activated at Fort Leavenworth, Kansas. Under the competent leadership of Colonels Edward Hatch and Benjamin Grierson, first Regimental Commanders of the 9th and 10th Cavalry Regiments, respectively, both regiments were trained and equipped and began a long and proud history.

For over two decades, the 9th and 10th Cavalry Regiments conducted campaigns against American Indian tribes on a Western Frontier that extended from Montana in the Northwest to Texas, New Mexico, and Arizona in the Southwest. They engaged in several skirmishes against such great Indian Chiefs as Victorio, Geronimo, and Nana. "Buffalo

Soldiers" was the name given the black cavalrymen resemblance between the black man's hair and the mane of a buffalo. Another view is that when a buffalo was wounded or cornered, it fought ferociously, displaying unusual stamina and courage. This was the same fighting spirit Indians saw in combat with black cavalrymen. Since Indians held the buffalo in such high regard, it was felt that the name was not given in contempt.

When not engaged in combat with Indians, both regiments built forts and roads, installed telegraph lines, located water holes, escorted wagon trains and cattle drives, rode "shotgun" on stagecoach and mail runs, and protected settlers from renegade Indians, outlaws, and Mexican revolutionaries. Elements of both regiments fought in Cuba during the War with Spain and participated in the famous charge on San Juan Jill. Troopers of the 10th Cavalry Regiment rode with General John J. Pershing during the Punitive Expedition in Mexico in search of Poncho Villa. In 1941, the two regiments formed the 4th Cavalry Brigade, commanded by General Benjamin O. Davis, Sr., at Camp Funston, Kansas. In 1944, the end came to the horse cavalry regiments and the curtain was lowered on the long and glorious past of "The Buffalo Soldiers."

Greater Los Angeles Area Chapter of the Ninth and Tenth (Horse) Cavalry Association

Evolution of The Calvary Association
By Trooper Turl Covington, Jr. and
Trooper Franklin J. Henderson

The 9th and 10th (Horse) Calvary Association traces its beginning to the year 1966 in Kansas City, Missouri. At that time, a group of former cavalrymen got together to talk about their military heritage and unique military experience. One hundred years had passed since, by an Act of Congress, two regiments of cavalry were created for colored men. Designated the 9th and 10th Cavalry, these regiments were part of a bold experiment to accept Black men in the regular Army establishment.

TROOPER - HAROLD S. COLE

At the Kansas City meeting were veterans of the 9th and 10th Cavalry Regiments. Many were assigned to the regiments when they were inactivated in March 1944. With their ranks getting thin, they looked for ways to assure that information about the exploits and accomplishments of the original Black cavalrymen and their own experience would not die with them. It was decided that this could be accomplished by forming a 9th and 10th Cavalry Association. It was also decided to hold annual reunions at different locations in the nation. Annual reunions, it was assumed, would attract the men who had served in the regiments and shared the common unique experience, would perpetuate the memory of comrades who have passed on, and would provide community awareness of their rich military heritage.

The 9th and 10th Cavalry Association was initially chartered in the State of Missouri as a non-profit organization. Later the name was changed to 9th and 10th (Horse) Cavalry Association to distinguish it from modern cavalry units. In 1985, the charter of the 9th and 10th (Horse) Cavalry Association was transferred from the State of Missouri to the State of Kansas.

At the outset, regular membership in the association was limited to persons who had served in the 9th and 10th Cavalry Regiments. Among the charter members were David Allen, Otto Bess, Albert Bly, Turl Covington, Jr., Charles Gates, William Harris, Earl Lewis, Jalester Linton, James P. Meigs, Jr., Nolan Self, Doyle Tate, and John E. Washington. In 1977, allied membership was extended to persons who had served in the 27th and 28th Cavalry Regiments. Later associate membership was extended to any person who has rendered outstanding service to the association or to the United States through service either in the Armed Forces or their community.

First day for Buffalo Soldier Stamp issued, April 22, 1994, Fort Huachuca, AZ

The first annual reunion, the 101st Anniversary Reunion, was held in 1967, at Fort Riley, Kansas. Since 1967, the association has held several successful annual reunions. Reunion guest speakers have include Colonel (later Lieutenant General) Julius Becton, Jr. General and Chairman, (Joint Chief of Staff) Colin L. Powell, Rear Admiral L.A.. Williams, Major General John Q.T. King, and Major General Harry Brooks.

An active effort of the Association is to stimulate public awareness and interest in the history and achievements of the Buffalo Soldiers. Most historians have overlooked or suppressed the role played by Buffalo Soldiers in the settlement and economic development of the western half of the United States after the Civil War. Moreover, their years of service to this nation, both at home and abroad, is just gradually becoming known. To increase public knowledge of their unique record of service, the Association will use its resources to engage in community activities that feature the contributions of black men and women to American military history.

HONORED TROOPER—COUNCILWOMAN RITA WALTERS INTRODUCES ARMY TROOPER CURTIS L. JAMES OF THE FAMED BUFFALO SOLDIERS TO THE LOW ANGELES CITY COUNCIL FRIDAY WHEN THAT BODY PRESENTED JAMES AS A CITY COMMENDATION FOR HIS ARMY SERVICE, JAMES CENTER, SERVED AS A ORDERLY TO GEN. B.O. DAVIS SR., THE FIRST BLACK GENERAL IN THE U.S. ARMY. DURING WORLD WAR II, JAMES WAS PART OF THE D-DAY LANDING IN NORTH AFRICA, AND THEN FOUGHT ON ANZIO, PALERMO AND SALERNO BEACHES IN THE INVITATION OF ITALY. HE HAS RECEIVED MORE THAN A DOZEN MEDALS AND RIBBONS. WITH HIM ARE HIS FELLOW BUFFALO SOLDIERS, BRUCE DENNIS, ROBERT PETTY AND FRED JONES, FROM LEFT. —FAREED MUWWAKKIL PHOTO.

9th Cavalry Regiment
- Constituted July 28, 1866 in the Regular Army as the 9th Cavalry
- Organized September 21, 1866 at Greenville, Louisiana
- Inactivated March 7, 1944 in North Africa
- Re-designated October 20, 1950 as the 509th Tank Battalion (Negro)
- Activated November 1, 1950 at Camp Polk, Louisiana
- Ordered Integrated December 1952

Trooper James Madison   Washington D. C. Ninth & Tenth (Horse) Cavalry Association

Campaign Participation Credit
- Indian Wars
- Spanish American War
- Philippine Insurrection
- World War II

10th Cavalry Regiment
- Constituted July 28, 1866 at the Regular Army as the 10th Cavalry
- Organized September 21, 1866 at Fort Leavenworth, Kansas
- Inactivated March 20, 1944 in North Africa
- Re-designated October 20, 1950 as the 510th Tank Battalion (Negro)
- Activated November 17, 1950 at Camp Polk, Louisiana
- Ordered Integrated December 1952

Campaign Participation Credit
- Indian Wars
- Spanish American War
- Philippine Insurrection
- World War II

Medal of Honor Recipients

9th Cavalry Regiment
- Captain Francis S. Dodge, Troop D
- 2nd Lieutenant George R. Burnett
- 2nd Lieutenant Robert E. Emmet, Troop G
- 2nd Lieutenant Matthias W. Day, CO I
- 1st Sergeant Moses Williams, Company I
- Sergeant Thomas Boyne, Company C
- Sergeant John Denny, Troop B
- Sergeant George Jordan, Company K
- Sergeant Henry Johnson, Company D
- Sergeant Thomas Shaw, Company K
- Sergeant Emanuel Stance, Company F
- Sergeant Brent Woods, Company B
- Corporal Clinton Greaves, Company C
- Corporal William O. Wilson, Company I
- Private Augusts Walley, Company I

10th Cavalry Regiment
- Captain Louis Powhatan H. Clarke
- Captain Louis H. Carpenter, CO H
- Sergeant Major Edward L. Baker, Jr.
- Sergeant William McBryar, Company K
- Private Dennis Bell, Troop H
- Private Fitz Lee, Troop M
- Private William H. Thompkins, Troop G
- Private George H. Wanton, Troop M

CHAPTER 3.

SERVICE TIME

Harold S. Cole at Fort Clark, Texas, 1943

As a young teenager in high school in 1939, there was a war in Europe. German submarines were torpedoing the U.S. shipping fleet in the Atlantic Ocean. On December 7, 1941, the Japanese Imperial Navy bombed Pearl Harbor. Two days later, the United States had declared war on Japan, Germany, and Italy.

Most young Americans were at the recruiting office in a short time to enlist in the military to defend their country. In 1942, I was one of those young Americans. My parents had to give their consent for me to sign up because I was seventeen years old. I was born and went to school in North Pelham and New Rocehell, New York. White and Black people did many things together. There was no segregation. So what happened when I entered the military? We were segregated. I managed to get over this shock because Black people are proud, and no matter what happens to us, we seem to get stronger and determined to be the best in whatever we do.

The military asked me what outfit I would like to join, infantry, engineer, or cavalry. When I was a young boy in North Pelham, two of the best people anyone had the pleasure to know were my neighbors. They were stationed at the United States Military Academy at West Point. Their names were Trooper Rollam Standsberry and his brother Trooper William Standsberry. There were cavalrymen and I admired them. I also had some experience horseback riding. My older brother, James B. Cole, worked at the Hutchinson Riding Academy and he rode horses every day. With this background, I told the military that I wanted to join the cavalry.

I was given a physical examination on White Hall Street in New York City, raised my right hand, and was sworn into the Army. I boarded a bus and was taken to a train station. I, along with other inductees, boarded a train headed for Camp Upton, Long Island, New York. We were not completely segregated, but I was assigned to an all-Black Company. White barbers gave the men in my Company a haircut, and we viewed a training film with White inductees. Black and White soldiers were in the same line for issue of clothing and equipment, and we had chow in the same mess hall. However Black and White inductees had separate quarters and training areas. I had been a Boy Scout. I knew how to march in formation and that was a plus. I hated to be told what to do. When to get up, when to eat, when to receive mail, when to stand retreat, and when to go on parade. A pass was needed to leave the installation and I was told what time to be back on the installation. It was hard for me to adjust to Army life. Issued clothing came in two sizes—too small and too large. Most soldiers tailored their clothing. You really don't know what talent you have until you join the Army. A soldier in my barracks had measles, so the entire building was quarantined. All we did for a month was eat, sleep, and drill on the parade grounds. We also played blackjack and each day someone new would win all the money. So the last man had big winnings.

I thought the train ride from New York City to Camp Upton took a long time. Fortunately, for me my good friend Henri A. Legendre accompanied me to our first duty station. Henri would go on to become a professional Architect and community leader. This initial trip was short compared to my next train ride. I boarded a train at Camp Upton headed for my next destination. I now know it was Fort Riley, Kansas, and from there to Fort Clark, Texas. A cross-country train ride at the time was another experience. Travel from New York to St. Louis was all right. After St. Louis, it was another world, one that I had not seen before. Black people in the rear and White people up front. Separate water fountains, toilets, cafe, stores, etc. The farther South the train traveled the more desolate it got.

I retrained at Spartanbury, Texas and was bused to Fort Clark in Brackettville, Texas. I was assigned to Troop F, 9th Calvary. I was very impressed with the way the troop functioned. Everyone was spit and polish. We were quartered on the old post, in good barracks, waiting for tarpapered-covered barracks to be built across the creek. It was not long before we moved. There was an awful lot of work to do along with training. Exercise and marching were daily routines. When I arrived at Fort Clark, there bad feelings between the troopers from New York and troopers from the Southern States. Main reason was the educated and the not-so-educated. But as time went by and training became more intense, we began to understand each other better. This led to a bonding that was great. Black troopers could not go to the PX, theater, or any facility on the old post except the post hospital. A put-together PX was placed in every troop. A Day Room was put together in each troop where beer and snacks were sold. Shortly thereafter, a theater and PX were built on our side of the post. During my service in the cavalry, strength played an important part in soldiering. The strong an intelligent troopers always got their way and were the first promoted and given special favors.

One day, the Platoon Sergeant told everyone to go to the corral and get a horse. He gave us a halter to bring the horse back to the stables. I entered the cavalry with about 500 troopers from New York. Many had never been on a horse or up close to one. For a small favor, I would get the horse they wanted. I always loved horses. In the cavalry, they are called "Big Eyes". Your horse had to be cleaned daily. We were issued brushes and curly combs, dock rags, and a pick. The brush and comb are used for the horse's fur coat and tail. The pick is used the clean the hooves. The dock rag is used to clean completely all areas under the tail. Your horse had to be clean at all times. At the end of the day, you would always water your horse before you fed it. If you fed the horse first, it would fill up with oats. Then when you give the horse water, it will drink until it was full and the oats in its stomach would swell and may kill the horse. My horse was name "Bob" after a pet dog that I had at home that died. Your horse was always checked by the Stable Sergeant and the Veterinarian. If the horse was sick for any reason, you did not ride it. You would have to get a horse off the picket line that belonged to no one. You would do that every day until your horse was well. Stable detail was performed by each platoon in a troop or a rotation basis. Stable hands were present, but to clean the stables every morning was a huge task. Therefore, a platoon would go to the stables at 0400 hours (4:00 am) to clean the troop stables. Stable cleanup would be finished at 0600 hours and the platoon would march to the troop mess hall for chow and drill call.

Weapons were assigned to us at the beginning of our training. I was issued a machete, bayonet, pistol (better known as a six shooter), and an 03 rifle (Springfield Model 1903). Later on, I was issued a caliber 0.45 automatic pistol and an M-1 rifle. Still later, I was issued a carbine. I had to learn how to disassemble and assemble all weapons while blindfolded. I was also issued a saddle, bridle, horse blanket, stirrups, gun boot, girth, halter, shelter half with pins, feed bag, saddle bags, spurs, poncho, clothing boots, breeches, pistol holster, web belt, magazine holder, cartridge belt, gas mask, steel helmet, helmet liner, helmet liner cap, gloves (three kinds), and a general issue of individual clothing. Our quarters were always spotless. I had a foot locker and a clothing rack for uniforms and hats. Shoes were lined up under my cot. As training continued, we troopers got better and became a good combat unit. I loved the cavalry and all it stood for.

Troopers in the mule pack were a very rough bunch, The biggest guys were picked to be mule skinners. They would raise hell every morning at 0400 hours when they were moving out to the field. I really admired them. We were allowed to ride our horses on Sunday and many troopers did so. The good side is that you did not have to ride in formation. We could ride at whatever gait we desired, jump our horses, and ride wherever we wanted to go. The bad side was that we had to feed our horses and clean our equipment. So if you ride every day, why would you want to ride on Sunday?

I did experience prejudice in my troop. I took and passed the test for Warrant Officer School. When it was time to enter the school, I asked the First Sergeant why I was not assigned to the class. He told me that none of his troopers were going to out-rank him. The funny thing about this situation is that the First Sergeant had also passed the same test.

In addition to my field training, I was a drummer in the Regimental Drum and Bugle Corps, Bugler for the 4th Brigade, 2nd Calvary Division, motion picture operator, and Platoon Sergeant. If Saturday was a free day, the unit would send a convoy of troopers to Mexico, if they desired to go, for rest and recreation (R&R). We always had a great time in Mexico. The people there loved us. I guess it was because we had money.

In 1943, the 2nd Cavalry Division was moved to North Africa. When we arrived in Oran, Algeria, the First Sergeant told everyone to dig a foxhole. We were on top of a mountain, nothing but rock below. All troopers griped about not having anything to dig into rock. So when night fell, there were very few foxholes. About 0200 hours, a German aircraft flew over Oran and bombed the troop area. The next day, I looked around at our area and to my amazement everyone had a huge foxhole. Like I said earlier, you never know what you can accomplish under adverse conditions.

In 1944, the War Department deactivated the 2nd Cavalry Division. It was very hard to accept this, but, as I said earlier, Black people do better when things are at their worst. Some troopers transferred to the 92nd Infantry Division. They wanted to fight the enemy. They were trained to fight. They were proud troopers and sad that their historic regiment was eliminated from active service while in a combat zone. Many troopers were ordered into service units. I was sent to a Port Battalion in Italy. I found that the White soldiers from the northern states were more prejudiced than those from the southern states. I participated in the invasion of Southern France. It wasn't long thereafter that the war in Europe ended. I was discharged in 1946.

CHAPTER 4.
MAKING HISTORY

Trooper Cole riding in one of the many parade ceremonies

Having the privilege of serving as the 18th President of the Ninth and Tenth (Horse) Cavalry Association was truly a prestigious honor and will serve as one of my fondest memories. Words cannot describe how proud I was to hold this position, and the seriousness I took to this position. By the time of my election, I was well prepared. I had previously served as chapter president, and knew very well the requirements and work that went into standing up a chapter. At the time of my tenure as President, we created more chapters across the nation than in any point since the founding of our great organization.

I wish I could take all the credit, but I can't! I was fortunate to be surrounded by gallant and faithful men and women who knew the proud shoulders they were standing. Because of the heroic heritage of our origin we were able to summon some very powerful and politically involved support from the highest office in the country. It is because of the collective dedication of the men and women of the Ninth and Tenth (Horse) Cavalry Association we were able to make huge strides.

Troopers Cole, Robert Powell, and William Woodell

I worked hard as a member of the Ninth and Tenth (Horse) Cavalry Association and decided to serve in several leadership positions; 1993—1995: First President and Co-Founder of the Greater Los Angeles Area Chapter of the 9th & 10th (Horse) Cavalry Association; 1993—1997: 18th 1st Vice President of the 9th & 10th (Horse) Cavalry Association; and 1997—2001: 18th President of the 9th & 10th (Horse) Cavalry Association.

I originally joined the Ninth and Tenth (Horse) Cavalry Association in 1989, and have been committed to its growth and expansion. As National President my goal was to travel to every chapter and communicate our national goals. One of the major accomplishments during my tenure was revising or updating the organization's Constitution and Bylaws. This was a tremendous undertaking and one that required every member's participation. The role of our organization is to maintain our tremendous heritage and honorably represent the legacy of Buffalo Soldiers of past and present. We have an obligation of etching and imprinting the significance of our

Trooper Cole attending one of the many meetings as National President

collective military service to everyone for as long as God grants the privilege to people to live.

More youth programs are needed to preserve our important legacy. One of our core principles is to stimulate patriotism in the minds of our youth by encouraging the study of the patriotic and military history of our nation. As technology continues to improve we have an obligation to update our educational platforms, and partner with other organizations with similar and like-minded goals and objectives. We need to rid ourselves from thinking we can have a positive impact across the nation, by going it alone. It requires support, and collaboration from various groups to meet this important demand. If educating our children is not a demand, then we need to make it one. Many times society may act in a fashion that illustrate their collective confusion that lacks leadership and vision. This is where the Ninth and Tenth (Horse) Cavalry Association shows up and do what we have always done. Even when the nation turned its back on us, the Black Soldier has always taken the high road and done what is right for this nation.

Trooper Harold S. Cole, as President of the Ninth and Tenth (Horse) Cavalry Association

We stand on the shoulders of some really proud Americans and we owe to their sacrifice never to forget, belittle, marginalize, or devalue their accomplishments, but to stand squarely on what is right for our nation, just as they did. One of the best initiatives we have in our organization is awarding scholarships to young people. By giving scholarships we ease the financial burden for parents, and create an opportunity for our youth to aspire to greatness. Our Ladies Auxiliary serves an essential role by assisting us successfully navigate this space. Women should always have an active role in the Ninth and Tenth (Horse) Cavalry Association.

When the Ninth and Tenth (Horse) Cavalry Association first started, members had to have served in the Ninth or Tenth Cavalry, but as time went on, we realized how limiting this was, and soon realized we would eventually all die off. We quickly changed this perspective and allowed others to join, and soon learned how impactful other people could be if given the opportunity to contribute value. Since that decision many of our members who came from outside of Army heritage became our biggest supporters and cheerleaders.

LTC Keith Bowle USA visits with members of the Greater Los Angeles Chapter

One such hero, is Trooper Derrick Davis, who came to us as a career long Atlanta Police Officer and gallantly served as a Mounted Police and K-9 Officer. Under his leadership we have made tremendous strides and will continue in making significant milestones. We need to never be limiting in our perspective, but always searching for new ways to improve society with the mindset of always looking back at where we have come from, and the tremendous ways God has blessed us.

Trooper Edwina Cole spending time in the saddle

It is vitally important that we continue to educate people on how to conduct yourself properly in life. There is nothing more important than obtaining your education. Before you can do anything you must have an education! The success of the Ninth and Tenth (Horse) Cavalry will reside in how well we educate our future. Additionally, one of our primary focus must be visiting the elderly. As Troopers continue to age, our organization has a responsibility to visit these pioneers who paved the way for all of us.

Throughout my career it has been my experience we have had to be twice as good, even if it takes twice as long in accomplishing an intended task. Racism has served as a leading cancer that is killing our country very slowly, and it will continue to ruin our nation unless deliberate policies are not implemented to prevent the unfair treatment of some. There is a dire need for radiation chemotherapy that offers a laser light focus on this phenomenon.

Family members supporting the Ninth and Tenth (Horse) Cavalry Association Reunion in Baltimore, MD

The creation of the Buffalo Soldiers was built because of racial beliefs, and in spite of these racial practices the Ninth and Tenth Cavalry answered the call for which it was created, by exceeding every requirement and mission given. Our organization is properly positioned to leverage our legacy to lead tough and sensitive conversations about resolving racism in this country. Combating racism will continue in being an emerging concern within our ranks as long as its heinous practices are in place in the United States of America.

Buffalo Soldier Monument
Fort Leavenworth, Kansas

CHAPTER 5.
FAMILY'S IMPACT

Sitting (L to R) Doris Cole, Cornelia Cole-Dance, George Leon Cole, Arline Bateman Cole, Evelyn Cole-Harris, Edith Cole-Robinson. Standing (L to R) Keith Cole, Donald Cole, Harold Cole, Andrew Cole, James Cole

The impact of the family, especially of the Black family is absolutely critical. I have been extremely fortunate have come from an amazingly loving and caring family. My father was an excellent care provider, and did the best he could with what he had. To ensure we had enough to support ourselves, everyone was expected to work, and contribute to the family. When someone in the family needed something, they would make it known, and if feasibly possible we would work as a unit in accomplishing the task. I feel really fortunate to have been raised in a loving family, because it shaped the person that I later became.

Trooper Cole in Naples, Italy in 1944

It was the teachings from my family that gave me the confidence to know I would do well in the Army. My brothers fully supported me and taught me how to ride and maintain a horse. All these life lessons shaped my thinking, and I would like to think gave me additional insights into the art of discipline, knowing how to listen, execute orders, and live in an organized way. I was never much on talking, but did when required, or relaxing with friends.

My family became the foundation for my learnings, and gave me a leg up in knowing what to expect out of life. You cannot live as long as I have and know experienced racial discrimination and prejudice. When growing up in New York, I lived amongst people from all over, and did not have to routinely endure the racism that went on in the South. It was normal to engage with Italians, Germans, Russians, Polish on a regular basis. These people were in the same financial situation as we were, everyone was trying to make a living. I did not really have to face Jim Crow racism until I joined the Army and was traveling to my first duty station.

Joining the military was a natural event for me. I loved riding horses, and my family had a tradition of serving. My father was a very special man who came home and loved his children deeply. Anything I ever wanted to talk to him about I could, and he would take time out for each of us. Like most families in these days, we did not live in excess and there was never a lot of waste. What little bit we had we took care of, and tried to preserve things for when we needed it.

Our neighbors were all decent people and did not create a ruckus. They too supported one another as best they could, as long as it did not diminish supporting their own family. I can recall moving around my community freely with no problems. Police did not harass us, and we lived like every other family in the neighborhood. When I was younger, my sisters were responsible for watching over us, and ensuring we completed our chores. Working in the Cole household was expected. No one ever got away without pulling their weight!

Growing up with nine children in the house there was always something going on! We really did not need a lot of neighborhood children to come around, because we had enough folks to field a team if we wanted to play stickball, kickball, hide-and-go-seek, or any other games. My sisters would normally be the ring leaders when it came to playing games.

Trooper Cole at Fort Clark, Texas - first duty assignment

My sisters and brothers and I were a very tight-knit group. My mother and father did not mind whopping us if we got out of line, but typically they did not have to, because one of my siblings would get to us first, and warn of the pending danger we were putting our self in. I must admit, I was a pretty good listener. It did not take me long to catch on to what was expected, and instead of seeking a lot of fanfare, I just did what I needed to do, and moved on. I learned this secret by watching my brothers, and older sisters.

Out of all the things I learned best, I believe it must have been about teamwork. I easily saw the importance of how working together and effectively communicating got things done. The great thing about having so many people in the house, is that you were never lonely. I guess, sometimes you may seek a little solitude, which was rarely found, but in my younger days I learned how to work together with other people. Even though this was my immediate family, these lessons would later transition me into a better military man.

Edwin, Leroy, Jeffrey, & Vicky Roberts

Growing up in our household and referring to our parents and other adults as "Yes Ma'am", and "No Sir" was just part of our culture. There were no calling adults by their first names. In school you were expected to respect your teachers, and not get out of line. I was a decent student; I guess I would consider myself as average. I made respectable grades, but was plenty distracted playing sports and chasing girls in high school.

High school for me was a lot of fun! I learned how to interact with people, the importance of follow through, and fulfilling what you said. People want to be around they can trust, and when you don't fulfill your promises, you will be considered untrustworthy. I was very athletic in high school and played baseball, basketball, and football. I must admit, I was pretty good and the coaches took notice of me. Being physically fit would play a significant role when I joined the Army.

Trooper Cole and Edwina Cole

Overall, I believe my upbringing and family life played a tremendous role in shaping who I became in life. Naturally, I have a lot of people to thank, and would be remiss if I did not highlight the importance of my family. Unfortunately, I waited to write this book until after my parents, brothers and sisters have departed, but they are live inside of me and there is not a day that I do not think about each of them. When I look at their children, I can see their faces, and remember the fondness of their presence. It is my prayer for the Cole decedents to recognize the greatness of their legacy, and do good in life by always remembering the shoulders they are standing. Their success did not just happen overnight, but was instilled deep inside of them throughout the ages, by their ancestors.

General Kip Ward presenting awards at the Ninth and Tenth (Horse) Cavalry Association Reunion

There are no self-made men and women, but any accomplishments we make are the results of the sacrifice and committed service of those who came before you. There are little original thoughts in life. What may be new to you, was once thought of, and conceived by someone before. Don't think too high of yourself, that you forget where you come from. Remain grounded and focused on what you want from life. Everything is possible as long as you keep your hand in God's hand and do not deviate from your
Spiritual beliefs

GREATER LAS VEGAS AREA CHAPTER

CHAPTER 6.
FORT CLARK

Horse Stables at Fort Clark, circa 1943

Fort Clark, Texas was established on June 20, 1852, as one of a chain of forts protecting the border with Mexico and the lower road to California. All but abandoned during the Civil War, by the early 1870's Fort Clark had grown to become the largest military installation in Texas. Generations of soldiers served at Fort Clark including the Buffalo soldiers of every historic black regiment in the Army, and the troopers of the Army's only black cavalry division.

On August 20, 1921, as a result of lessons learned from World War I, the US Army Adjutant General constituted the 1st and 2nd Cavalry Divisions to meet future mobilization requirements. As organized, 2nd Cavalry Division was to be the Army's only integrated division.

Placed on the rolls of the Army in 1921, the 2nd Cavalry Division was not activated until April 1941. As part of the Protective Mobilization Plan, the division was reserved for activation at Fort Riley, Kansas, but due to manpower constraints it never reached full strength. The 2nd receive the appropriate number of cavalry regiments, but units providing the organic support and service troops remained unfilled. The first cavalry regiments, but units providing the organic support and service troops remained unfilled. The first divisional activations came in October 1940, with the organization of the 3rd Cavalry Brigade and the assignment of the 2nd Cavalry Regiment (served at Fort Clark 1856 – 1859) and 14th Cavalry Regiment (served at Fort Clark 1912 – 1916). The 4th Cavalry Brigade activated during February 1941, with the 9th Cavalry Regiment (served at Fort Clark 1876 – 1873; 1875 – 1876; and 1899 – 1900) and 10th Cavalry Regiment (served at Fort Clark 1876 – 1878 and 1900 – 1901) as its cavalry regiments. These last two regiments, the only two available for assignment, were the framed Buffalo Soldier units of the Indian Wars era. The division, therefore, was unique to Army structure at that time, a racially mixed unit.

Split between Fort Riley and Camp Funston, Kansas, neither post having adequate facilities for the division's horse cavalry, personnel shortages continued and divisional elements were activated using provisional assets. General Millikan, the 2nd Cavalry Division Commander in 1941, envisioned a combined use of mechanized and horse cavalry within the division. During July, Troop A, 2nd Reconnaissance Squadron, was formed provisionally as a mechanized divisional element. The division, now organized with horses, scout cars, jeeps and motorcycles, spent most of the summer training with its new equipment.

Fort Clark Gideon 1943

During the spring of 1942, a War Department decision to increase the number of armored divisions within the United States Army resulted in the planned conversion of the 2nd Cavalry Division. White troops in the 3rd Brigade were used in the formation of the 9th Armored Division. The 2nd and 14th Cavalry were inactivated and their personnel transferred into the newly formed 2nd and 14th Armored Regiments, both elements of the new armored division. On 15 July 1942, the 2nd Cavalry Division was inactivated. The 4th Cavalry Brigade with its black regiments, however, remained active. The activation of the 9th Armored Division created logistical problems at Fort Riley and Camp Funston. The installations that had accommodated a single division were now home to a division and an additional cavalry brigade. Consequently, the 4th Cavalry Brigade Headquarters and the 10th Cavalry, relocated to Camp Lockett, California. To provide cadre for the 2nd Cavalry Division, the Buffalo Soldiers of the 9th Cavalry returned to Fort Clark in November 1942 where the regiment had last served in 1900.

Boxing Champion Sugar Ray Robinson

As the number of Black personnel entering the Army rose, the need for units for these soldiers to join also increased. In November 1942, the War Department directed that the 2nd Cavalry Division would be reactivated, and that two new Black regiments would be assigned. It was also announced that the 2nd, now the Army's third Black division, would remain divided between Texas and California. Construction was started at both posts since neither had the facilities to support an entire division. The work completed, the 2nd Cavalry Division activated on 25 February 1943 with Headquarters at Fort Clark. The 9th and 27th Cavalry, active at the Texas post, were the assigned troops of the 5th Cavalry Brigade with over 10,000 soldiers ultimately stationed at Fort Clark. The 10th and 28th Cavalry, located at Camp Locket, made up the 4th Cavalry Brigade.

Command of the Division fell to Major General Harry H. Johnson, native Texan and first cousin to future President Lyndon Baines Johnson, and also considered by many at the time to be second only to George S. Patton, Jr. as the toughest cavalryman in the Army.

SGT Joe Lewis, 1943

Filled using recruits straight from the induction centers, the 2nd Cavalry Division spent most of the spring and summer of 1943, at Fort Clark training its soldiers. The division provided these men with their basic training as well as instruction in Cavalry operations. The 2nd Cavalry Division was the largest mounted formation ever stationed in Texas with over 5,000 horses. In the early months of 1944 the days of the horse in the Army came to an end and the 9th and 27th Cavalry Regiments were the last organization in the Army to turn-in their horses.... forever!

The massive building program at Fort Clark to accommodate the 2nd Cavalry Division included several hundred temporary frame and tarpapered barracks which literally baked in the scorching Texas sun, and duplication of several existing facilities in order to separate Black soldiers from White soldiers. A second service club, second theater, and a second swimming pool were all hastily constructed far from the main post in the troop cantonment area east of Las Moras Creek. These amenities, along with organized athletics and U.S. Special Services events provided welcome relief from the stress of the Division's relentless training program.

A special highlight of the 2nd Cavalry Division's service occurred in June 1943, when 20th Century Fox Movietone News featured the Division and Fort Clark in their newsreel the "Negro Cavalry Unit."

In December 1943, Army Special Services visited Fort Clark with an entourage of boxers which included heavyweight champion of the world Staff Sergeant Joe Lewis and Corporal Walker Smith (Sugar Ray Robinson). Exhibition bouts drew as many as three thousand spectators. I was one of those spectators cheering on the boxers! On December 3rd Joe Louis fought Private Ezzard Charles of the 77th Field Artillery in one of these exhibitions. After the war, Charles fought professionally and inherited the heavyweight crown when Louis retired in 1950. Of additional significance is the brief service of 2nd Lieutenant Jackie Robinson, future baseball hall of famer for the Brooklyn Dodgers, with the 27th Cavalry at Fort Clark.

Second Cavalry Division Patch

The 2nd Cavalry Division's time at Fort Clark remains shrouded in controversy and rumor fueled by unsubstantiated stories of murder, riot, and conflict brought on by strained race relations, so prevalent in American society and the military at the time. In July 1943, Brigadier General Benjamin O. Davis, Sr. the Army's first Black General Officer, spent six days at Fort Clark and then wrote a report to the War Department's Inspector General. He concluded that "there is a feeling of resentment against the White Officers as a group by many colored officers and enlisted men." He called for removal of a White Lieutenant Colonel in the 9th Cavalry, and desegregation for Black Officers in messes and the Post Theater. He also reported that the Division Commander didn't seem to take Black soldiers' complaints seriously. Veteran troopers of the 2nd Cavalry Division, who returned to Fort Clark in September 2008 after a sixty-five-year absence, recalled the racial injustices of the times but preferred the memories of passes to San Antonio and Del Rio and the ice cream at the post beer garden to any now long forgotten ill-feelings they may have held towards Fort Clark.

The division's combat readiness would not be tested. Stating that there was no intrinsic need for a second cavalry division, the War Department had devised a plan to use the 2nd Cavalry Division personnel to form needed service units.

One day in late December 1943, word wet out on the post for all personnel to listen to a particular radio program. The voice came on the air, "Ladies and Gentlemen, welcome to a program of band music from New York, this is Fred Waring speaking. This afternoon we are going to salute General Harry Johnson and the officers and the men of the Second Cavalry Division down at Fort Clark, Texas. General Johnson's troops have been undergoing intensive training since last February and we show our appreciation with the following number." Thereupon the band, Waring's Pennsylvanians, swung into a snappy march rendition of "Stouthearted Men." This was the tip-off. The Division's orders for overseas shipment followed expeditiously thereafter.

Black community leaders, reacting against the criticism of the performance of blacks in combat units, protested the possible conversion of the division. The debate over the capabilities of Black units continued but the decision concerning the status of the 2nd Cavalry Division was already made. The War Department ordered the division to be shipped overseas where the conversion would take place. During January 1944 the 2nd Cavalry Division was dismounted and shipped back east for deployment abroad. Arriving at Oran, North Africa on March 9, 1944, the division was inactivated the next day.

Trooper Cole takes time out for riding with the Las Vegas Chapter

Only months after the departure of the 2nd Cavalry Division Fort Clark met a similar fate, slipping quietly and unceremoniously into history and out of active service on August 28, 1944, when the last soldiers departed. Even before that, into the fall of 1945, German POW's were engaged in dismantling buildings. The vulnerable post was declared surplus and sold for salvage in October of 1946. The new owners tore down the 2nd Cavalry Division cantonment area's nearly 1500 wood frame buildings. The Brown Foundation operated the grounds as the Fort Clark Guest Ranch until 1971 when the property was sold to a private developer who created a gated community and homeowners association, which operates today as the Fort Clark Springs Association.

Ninth Cavalry Honor Guard

Thus the final chapter of the Buffalo Soldier legacy in Texas was written at Fort Clark in 1943-1944, by the troopers of the 2nd Cavalry Division. Tanks and horses trained side by side in the only Black Cavalry Division in U.S. Army history. Thousands of Black men answered the Nation's call to arms in World War II and served with honor far from home at a Texas post steeped in the proud history of the Buffalo Soldiers. Aged veterans of service with the 2nd Cavalry Division returning today to Fort Clark in the twilight of their lives can still hear the faint bugle calls and smell the thousands of horses which are the faded icons of their youth.

Congresswoman Maxine Waters (D) California has been a long time active supporter of the Ninth and Tenth (Horse) Cavalry Association and preserving our history by correcting military records in a just manner.

CHAPTER 7.
FORT SILL

Lt Henry O. Flipper, first Black graduate of West Point, and first Black officer assigned to Fort Sill, January 1, 1878

The site of Fort Sill was staked out on January 8, 1869, by Major General Philip H. Sheridan who led a campaign into Indian Territory to stop hostile tribes from raiding border settlements in Texas and Kansas.

Sheridan's massive winter campaign involved six cavalry regiments accompanied by frontier scouts such as "Buffalo Bill" Cody, "Wild Bill" Hickok, Ben Clark and Jack Stillwell. Troops camped at the location of the new fort included the 7th Cavalry, the 19th Kansas Volunteers and the 10th Cavalry, a distinguished unit of Black "Buffalo Soldiers" who constructed many of the stone buildings still surrounding the old post quadrangle.

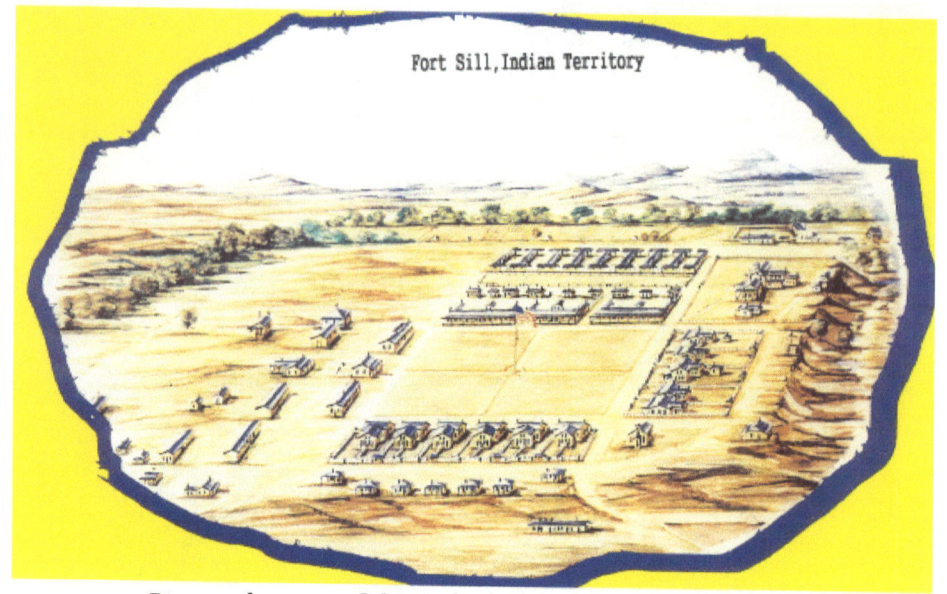

Picture diagram of the early days of Fort Sill, circa 1870s

At first the garrison was called "Camp Wichita" and referred to by the Indians as "The Soldier House at Medicine Bluffs." Sheridan later named it in honor of his West Point classmate and friend, Brig. General Joshua W. Sill, who was killed during the Civil War.

The first post commander was Brevet Major General Benjamin Grierson and the first Indian agent was Colonel Albert Gallatin Boone, grandson of Daniel Boone.

Peace Policy. Several months after the establishment of Fort Sill, President Grant approved a peace policy placing responsibility for the Southwest tribes under Quaker Indian agents. Fort Sill soldiers were restricted from taking punitive action against the Indians who interpreted this as a sign of weakness. They resumed raiding the Texas frontier and used Fort Sill as a sanctuary. In 1871, General of the Army William Tecumseh Sherman arrived at Fort Sill to find several Kiowa chiefs boasting about a wagon train massacre.

When Sherman ordered their arrest during a meeting on Grierson's porch, two of the Indians attempted to assassinate him. In memory of the event, the Commanding General's quarters were dubbed Sherman House.

Red River Campaign. In June 1874, the Comanche, Kiowa and Southern Cheyenne went on the warpath and the South Plains shook with the hoof beats of Indian raiders. The resulting Red River Campaign, which lasted a year, was a war of attrition involving relentless pursuit by converging military columns.

Without a chance to graze their livestock and faced with a disappearance of the great buffalo herds, the hostile tribes eventually surrendered. Quanah Parker and his Quohada Comanche were the last to abandon the struggle and their arrival at Fort Sill in June 1875 marked the end of Indian warfare on the South Plains.

Until the territory opened for settlement, Fort Sill's mission became one of law enforcement and soldiers protected the Indians from outlaws, squatters and cattle rustlers.

Geronimo. In 1894 Geronimo and 341 other Apache prisoners of war were brought to Fort Sill where they lived in villages on the range. Geronimo was granted permission to travel for a while with Pawnee Bill's Wild West Show and he visited President Theodore Roosevelt before dying here of pneumonia in 1909. The rest of the Apaches remained on Fort Sill until 1913 and they were taught by Lt. Hugh L. Scott to build houses, raise crops and herd cattle. Scott also commanded Troop L of the 7th Cavalry, a unit comprised entirely of Indians and considered one of the best in the west. Indian scout I-See-O and other members of the troop are credited with helping tribes on the South Plains to avert the bloody Ghost Dance uprising of the 1890 in which many died to the north.

Original Buffalo Soldiers (L to R) Turl Covington, Ronnie Hargrove, John Wright, Lonnett H. Eypert, Eugene Lewis, Merrill R. Anderson (wheel chair)

The Frontier Disappears. The last Indian lands in Oklahoma opened for settlement in 1901, and 29,000 homesteaders registered at Fort Sill during July for the land lottery. On August 6, the town of Lawton sprung up and quickly grew to become the third largest city in Oklahoma.

With the disappearance of the frontier, the mission of Fort Sill gradually changed from cavalry to field artillery. The first artillery battery arrived at Fort Sill in 1902, and the last cavalry regiment departed in May 1907.

The School of Fire for the Field Artillery was founded at Fort Sill in 1911, and continues to operate today as the world-renown U.S. Army Field Artillery School. At various times Fort Sill has also served as home to the Infantry School of Musketry, the School for Aerial Observers, the Air Service Flying School, and the Army Aviation School.

Troopers Cole and Washington visiting the Buffalo Soldier Monument in Junction City, Kansas

Today as the U.S. Army Field Artillery Center, Fort Sill remains the only active Army installation of all the fort on the South Plains built during the Indian wars. It served as a national historic landmark and home of the Field Artillery for the free world.

Ninth Cavalry Coat of Arms

"We Can; We Will!"

Tenth Cavalry Coat of Arms

"Ready and Forward"

Plaque at the US Military Academy

This athletic field is named in honor of the men of the 9th and 10th U.S. Cavalry Regiments, detachments of which once served at West Point. These Regiments of Horse Cavalry were first created by the Army Reorganization Act of 1866, and their early service was on the western frontier. They were composed of African American Troopers, who were called "Buffalo Soldiers" by their Indian foes, a sobriquet they adopted with pride.

During the Indian Wars of 1867 – 1891, the 9th and 10th Cavalry participated in 11 campaigns against hostile Indians, against Kiowa, Comanche, Utes, Cheyenne, Arapahoe's, Kickapoos, Apaches and Sioux. They were engaged in over 125 recorded battles and skirmishes, most taking place in Texas and New Mexico, but also the Dakotas, Idaho, Montana and Mexico. Some were major engagements, but many were detachment actions in which non-commissioned officers held command. During these actions there were many examples of hardships withstood and heroism displayed. Other Buffalo Soldier duties included guarding the border, apprehending bandits and cattle thieves and maintaining order in a sparsely settled and unruly territories.

In the Spanish American War, both Regiments were in the Cuban Expedition of 1898. The 10th made the frontal attack in the opening engagement of Las Guasimas and both Regiments participated in the attack on San Juan Hill. The 10th extricated the 'Rough Riders' from difficulty and then joined them in the assault on the blockhouse. Both regiments were also engaged in the siege of Santiago. Subsequently, the 9th Cavalry was sent to the Philippines, where they say action during numerous skirmishes from 1900 – 1902, during the insurrection. The 10th eventually returned to border duty in the southwest and accompanied General Pershing, during the punitive Expedition of 1916, engaged at Agua Caliente, Parral and Carrizal.

In 1907, a detachment of the 9th Cavalry was assigned to West Point and taught equestrian skills and gave mounted drill instruction. These skills were taught on the grounds now called, Buffalo Soldier Field, formerly known as Cavalry Plain. In 1931, the 9th detachment was replaced by the 2nd Squadron of the 10th Cavalry, which remained at West Point until 1941, when the Squadron rejoined the parent unit in Kansas.

Greater Los Angeles Chapter

General Colin Powell took a personal interest in seeing the creation and construction of the Buffalo Soldier's Monument at Fort Leavenworth, Kansas. He continues to lead and offer inspiration over the years.

92nd Infantry Division

24th Infantry Division Shoulder Patch

CHAPTER 8.   THE 24TH U.S. INFANTRY REGIMENT

Members of the 24th Infantry Division marching in formation in a community parade

The 24th Infantry's true history has its roots in the Constitution of the United States. By an Act of Congress, dated 28th July 1866, the 38th and 41st Infantry Regiments were created. These units were activated and manned with Black troops. A little more than 3 years later, on November 1, 1869, the 38th and 41st regiments were consolidated into a single combined unit – the 24th Infantry Regiment. The military ceremony for this occasion took place at Fort McKavitt, Texas.

In 1880, the 24th Infantry was sent into action in Indian territory. There it participated in a campaign against the Shoshonean Comanches. For a period of 10 years, the 24th served with great distinction no the Western frontier in many engagements with the Indians.

In 1898, during the Spanish-American War in Cuba, the 24th Infantry Regiment again distinguished itself when it victoriously participated in the Battle of San Juan Hill as part of the 3rd Brigade, 1st Infantry Division. It was this action, the storming of the infamous San Juan Hill blockhouse at Santiago, that inspired the design of the regimental insignia. This Blockhouse insignia bears the words "San Juan," as well as, the Latin words: "semper Paratus" (Always Ready) – the regimental motto. Upon cessation of hostilities, the 24th played a heroic part in the fight against the yellow fever epidemic at Siboney, Cuba. There many of the Black soldiers gave their lives in this great humanitarian effort.

During the years from 1899 to 1913, the 24th was sent to the Philippine Islands three times to help quell insurgents. Back in the United States in the spring of 1916, the 24th Infantry was called upon again. This time the regiment participated in the punitive expedition into Mexico against Poncho Villa and his forces.

For 20-year period, between 1922 and 1942, the 24th Infantry was stationed at Fort Benning, GA, where it was assigned to the Army Infantry Training School.

In April of 1942, the 24th Infantry sailed from the United States to the New Hebrides and Guadalcanal where they became the first Black combat troops to enter the Pacific theater of operations. Once there, they fought their way up the Solomon's, the Russel's, and the Marianas while heading to their main objective – Okinawa, "Keystone" of the Pacific. During this period, in early 1945, Colonel Julian G. Hear, Jr., Commander of the 24th Regiment, received the first ever formal surrender of a Japanese garrison to American troops.

Fellow Buffalo Soldiers enjoying quality time together

The 24th went on to Okinawa. While there, the war ended. In fact, it was soldiers of the 24th Infantry Regiment that handled security duties for Japanese peace emissaries who traveled to Okinawa in the "green-crossed" plane to plan the final surrender of Japan in 1945.

In 1946, while garrisoned on the island of Ie-Shima, the 24th Regiment joined the 25th Infantry "Tropical Lightning" Division in Japan to serve as occupation troops. The 3rd Battalion of the 24th was the first unit to join the 25th Division in October. The main body of the regiment came up from Ie-Shima in February of 1947.

Past-President Trooper COL Franklin J. Henderson presenting an award to Trooper Cole

The regiment received very high commendations for maintenance of order during the Kobe City Korean riots in April, 1948. This was Japan's only limited emergency during its occupation. The regiment was again commended for relief work and assistance provided following the Fukui earthquake disaster in June, 1948.

On June 25, 1950, following the outbreak of hostilities in Korea, the 24th Regiment was once more called upon to fight. By July 12, 1950, the 24th was committed to action as part of the 25th Division. Throughout 14 and one half months of extended frontline duty, the 24th Infantry served as a most valiant military unit.

Trooper Turl Covington, Jr. has served as a Past National President, Co-Founder, Life Member, Troop "A" 10th Cavalry, Headquarters Troop, 2nd Cavalry Division, Denver, Colorado. A leader, mentor, and great friend!

From day one, November 1, 1869, at Fort McKavitt, Texas, when the regiment was first activated as a single combined unit, until it was deactivated at Chipori, Korea, on October 1, 1951, one month short of its 82nd birthday, the 24th Infantry Regiment has served its country with honor, pride and professionalism.

In August, 1995, the 1st Battalion of the 24th Regiment was reactivated at Fort Lewis, Washington.

```
              UNITED STATES ARMY FORCES
                 PACIFIC OCEAN AREAS

                      COMMENDATION

                  The Commanding General
          United States Army Forces, Pacific Ocean Areas
                         Commends
                 The 24th Infantry Regiment
             for Exceptional and Meritorious Service
                  in the war against Japan

        From 20 December 1944 to 19 June 1945 members of the 24th
        Infantry Regiment, charged with elimination of the Japanese
        remaining on Saipan Island after organized resistance had
        ceased, maintained their organization's long and distin-
        guished history in the face of a determined defense by a
        fanatical enemy. Despite hardships engendered by the diffi-
        cult terrain and climate and the necessity for unceasing
        vigilance, casualties were held to a minimum and an unusu-
        ally high state of morale existed. The military profi-
        ciency and unwavering devotion to duty displayed at all
        times by the 24th Infantry Regiment were material contri-
        butions to the success of our forces in the Pacific Ocean
        Areas.
                                        /s/ Robert C. Richardson Jr
           (SEAL)                           United States Army

           A TRUE COPY:

                   GROVER C. ROSE JR.,
                   1st Lt, 24th Inf
                   Adjutant.
```

The above letter is extremely difficult to read, and is an official correspondence signed by LT General Robert C. Richardson, Jr., Comander, 24th Infantry., US Army. The citation reads; "From 20 December 1944 to 19 June 1945 members of the 24th Infantry Regiment, charged with elimination of the Japanese remaining on Saipan Island after organized resistance had ceased, maintained their organization's long and distinguished history in the face of a determined defense by a fanatical enemy. Despite hardships engendered by the difficult terrain and climate and the necessity for unceasing vigilance, casualties were held to a minimum and an unusually high state of morale existed.

CHAPTER 9. THE 25TH U.S. INFANTRY REGIMENT

Troopers taking time out for a quick picture

The military proficiency and unwavering devotion to duty displayed at all times by the 24th Infantry Regiment were material contributions to the success of our forces in the Pacific Ocean Areas.." This memorandum is dated June 2, 1945. The 24th Infantry received several letters of recognition for their bravery and outstanding service during their existence.

The 25th Infantry Regiment received its start very similar and under the same conditions as the 24th Infantry Regiment. The 25th Infantry would serve gallantly from 1866 to 1947, seeing action in the American Indian Wars, Spanish American War, Philippine American War, and World War II. After the Civil War, a reduction in military strength was conducted downsizing from forty-five infantry regiments down to twenty-five, with two regiments reserved for Colored Troops.

The 39th and 40th Regiments were consolidated and renumbered as the 25th Infantry Regiment. In April 1869, the set up their headquarters at Jackson Barracks, Louisiana.

In May 1870, the 25th Infantry was deployed to San Antonio, Texas, and after a short period smaller companies were distributed across several small Texas outposts, to include Fort Clark, Fort Bliss, Fort Davis, and Fort Stockton. These Buffalo Soldiers were sent to several cities on the Mexican border in Texas and New Mexico for the next ten years, providing border security, installing telegraph lines, building roads, and repelling the Indians.

In 1880, the 25th was transferred to monitor the northern Great Plains, in Montana, Minnesota, and the Dakota Territory. Their primary mission was fighting the Indians, and creating safe passage for pioneers moving west. Even though many of the Western movies, and later television shows would never highlight the fact the role of the Buffalo Soldiers, but they are the reason the West was tamed.

In 1894, the 25th was called on to provide protection during the railroad strike. In 1896 – 1897, they were asked to test improving the mobility for troops on bicycles. This research resulted in a summer-long bicycle ride from Fort Missoula, Montana, to St. Louis, Missouri, which is just shy of 2,000 miles.

Trooper Paul J. Matthews, Commander of the Buffalo Soldiers National Museum, 3816 Caroline St., Houston, TX 77004

Troopers Harold and Donald Cole

In gearing up for the Spanish American War in 1898, all four Color Regiments were deployed to Florida, and sent to Cuba to aid in rescuing the Rough Riders. Today, many members of the 24th and 25th Regiment are active members of the Ninth and Tenth (Horse) Cavalry Association. Some things just don't change. We fought together back in the late 1800s and we continue to support one another now.

At Fort Brown, Texas in summer of 1906, several men of the first battalion were accused of shooting Brownsville civilians and killing a bartender. President Theodore Roosevelt attributed the failure to prove soldier culpability for the attack to a conspiracy of silence and dismissed without honor all 167 men in the battalion, but fourteen were later allowed to reenlist.

First Sergeant Mingo Sanders, who had twenty-five years of service and had fought in Cuba and the Philippines was the one who lost everything, and became the symbol of injustice by the President.

During World War I, the 25th would become part of the 93rd Infantry Division, and would deploy to Champagne, Verdun, Aise, and Anould. The unit was demobilized in March 1919, then reactivated at Fort Huachuca, Arizona, in March 1942. The unit was inactivated at Fort Benning, Georgia in 1947.

LADIES AUXILIARY OFFICERS

President
Josephine Robinson

1st Vice President
Consuelo Curley

2nd Vice President
Iris Reddick

Secretary
Rhonda Rhodes

Treasurer
Doris B. Henderson

Asst. Treasurer
Marilee Scarbrough

Chaplain
Elizabeth Thompson

Historian
Louise Mikell

Parliamentarian
Jodie Morris

PHOTO NOT AVAILABLE

Sergeant-At-Arms
Debra Streets

Budget / Audit
Edith Brown

Budget/Audit
Betty Bruton

Scholarship
May Samuels

Scholarship
Annie Jones-Gray

Budget / Audit
Marva Matthews

25th Infantry Division Shoulder Patch

Ninth and Tenth (Horse) Cavalry Association Patch

CHAPTER 10.
REMOUNT

Los Angeles Mayor Tom Bradley and Trooper Harold S. Cole

The Poet Langston Hughes pinned the poem "Mother to Son" where a mother shares with her son, "Well, son, I'll tell you, life ain't been no crystal stair". Like this poem life for me has not been a crystal stair, and it had tremendous highs and very deep lows. As I reflect back I conclude, that the Lord has been mighty good to me and he has blessed and kept me. I never thought for one minute that being introduced to the military at a very young age would end up being the center piece of my life.

I have been able to make good by the military, by learning leadership, discipline, the meaning of hard work, integrity, and most importantly learning how to care for people. For all the right reasons, serving in the Army elevated my life, by giving me new experiences, hopes and dreams. I was able to complete a twenty-year career by serving in both the Army and Air Force.

Trooper Cole visiting the Civil Rights Monument

The jobs I held after retirement, was largely due to the experiences I learned while on active duty. Most of my best friends were met through the military, and naturally the Ninth and Tenth (Horse) Cavalry Association. For their friendship and loyalty, I say thank you! I will never really be able to thank you enough for the support, friendship, and positive difference we are making in communities.

I decided to write this book for several reasons. Primarily, because I am the last sibling living in my family, and I feel I need to leave a legacy, and secondly, no one else can tell my story but me. There is a story inside of all of us, and we owe the next generation the opportunity to learn from those who came before. Some may think, "well I never really did anything too significant?" But it is my assessment, if the Lord allowed you to live for over ninety years, you have a story to tell. Actually, I wish I had written this book sooner!

We all have something to give, and this is my way of highlighting my life's experiences by saying thank you to everyone who have played an active role in shaping me into the man that I am. Many may say, that I am a bit rough in my delivery, and they are right! I don't like wasting time on things that really don't matter, or can serve as a distraction. I have always been a man of principle, and did what I thought needed to be done.

In the course of my military career I spent four years in the Army and sixteen years in the Air Force. My career was not a typical career, but it was a full one. There were times, I was not treated fairly, but I overcame by enduring the nonsense of life by not quitting. The Good Lord has blessed me far more than I deserved. I have always prided myself on my ability to learn technical skills quickly. This trait served me both inside the military as well as a civilian. I continue being a veracious reader to this day. I just like knowing stuff. Reading has always been the logical way of improving your position in life. When you know, you know! When you don't know is where you will find trouble and confusion.

When President Barrack Obama was elected, it brought tears to my eyes. I just never thought we would ever elect a Black President, at least not in my lifetime. I am so proud of him for all he has done, and all that he stands for. I want so desperately for all young people to focus on getting a good education and paying close attention in school. It just does not make sense for young people to have the opportunity to learn, and then they squander it away in foolishness!

Harold S. Cole in 1972 and 2003. Still good-looking and distinguished!

Don't get me wrong, I spent ample time drinking, partying, and having fun, but when it came to working, I never compromised my working conditions by seeking the wrong things. I believe the key to success is surrounding yourself with the right people. I am not necessarily talking about religious people, or "do-gooders" but good honest people who care about your wellbeing and success. Life is far too short to have it any other way.

Take time out and spend quality time with family. The Creator only gave you one family, and you need to enjoy them while you have a chance, regardless of how crazy they may seem at times. I have loved and adored my family and we went through thick and thin, a lot of good and bad times, but we endured. I am now being given the treasure of being the last one from my immediate family and currently serve as our patriarch. This is a position I do not take lightly, because I know my days are numbered. I hope during my later years I have served as a living example of modeling good values and principles, and that my family decedents can look at my life as one of service, respect, and love.

Leadership is not a position to be sought, but earned. I have worked for both good and bad leaders, and I know what I speak. The collective experiences gained over nine decades of life, has proven to me that you have to follow your gut-instinct. This is your "still small voice" speaking sense to you. Learn to listen to it, and it may take a little training, but if you close your mouth, and open your ears your spirit will speak loudly to you. Regardless of the trouble you might be in, or the difficulties you may see, all the answers you ever need are deeply planted inside of you. Close yourself off from the outside and find some quite time to simply listen.

Well, son, I'll tell you:
Life for me ain't been no crystal stair.
It's had tacks in it,
And splinters,
And boards torn up,
And places with no carpet on the floor --
Bare.
But all the time
I'se been a-climbin' on,
And reachin' landin's,
And turnin' corners,
And sometimes goin' in the dark
Where there ain't been no light.
So boy, don't you turn back.
Don't you set down on the steps
'Cause you finds it's kinder hard.
Don't you fall now --
For I'se still goin', honey,
I'se still climbin',
And life for me ain't been no crystal stair.
Langston Hughes' Mother to Son

All the best,

THE WHITE HOUSE
WASHINGTON

Happy birthday! We extend the thanks of a grateful Nation as you celebrate this special occasion.

Our veterans represent what is best about this country, and each day, we see the enduring spirit of America reflected in their service. Throughout our history, we have been blessed with an unbroken chain of patriots who have come forward to defend that powerful promise we hold so dear – life, liberty, and the pursuit of happiness.

We wish you all the best in the years ahead.

Sincerely,

MEDAL OF HONOR RECIPIENTS

Black soldiers have demonstrated their valor, courage, and patriotism in each of America's wars. Their achievement have been so profound as to merit the award of our nation's highest and most revered award—The Medal of Honor. Below, are the names of those so courageous who were awarded the Medal of Honor. Sergeant William Carney was the first Black Medal of Honor winner in the Civil War. Truly, I salute their service and commitment to our freedoms.

The American Civil War

SGT William H. Carney, Co C, 54th Mass Colored Infantry
PVT William H. Barnes. Co C, 38th US Colored Troops (USCT)
1SG Powhatan Beaty, Co G, 5th USCT
1SG James H. Bronson, Co D. 5th USCT
SGM Christian A. Fleetwood, 4th USCT
PVT James Gardiner, Co I, 36th USCT
SGT James H. Harris, Co B, 38th USCT
SGT Alfred B. Hilton, Co H, 4th USCT
SGM Milton M. Holland, 5th USCT
CPL Miles James, Co B. 36th USCT
1SG Alexander Kelly, Co F, 6th USCT
1SG Robert Pinn, Co I, 5th USCT
1SG Edward Radcliff. Co C, 38th USCT
PVT Charles Veal. Co D. 4th USCT
SGM Thomas Hawkins, 6th USCT
SGT Decatur Dorsey, Co B. 39th USCT

Indian Wars 1870 – 1890

SGT Emanuel Stance, Troop F, 9th Cavalry
PVT Adam Paine, Seminole Indian Scouts
SGT John Ward, 24th Infantry Indian Scouts
Trumpeter Isaac Payne, Indian Scouts
PVT Pompey Factor, Indian Scouts
CPL Clinton Greaves, Troop C, 9th Cavalry
SGT Thomas Boyne, Troop C, 9th Cavalry
SGT John Denny, Troop B, 9th Cavalry
SGT Henry Johnson, Troop D, 9th Cavalry
SGT George Jordan, Troop K, 9th Cavalry
SGT Thomas Shaw, Troop K, 9th Cavalry
1SG Moses Williams, Troop I, 9th Cavalry
PVT Augustus Walley, Troop I, 9th Cavalry
SGT Brent Woods, Troop B, 9th Cavalry
SGT Benjamin Brown, Company C, 24th Infantry

CPL Isaiah Mays, Company B, 24th Infantry
CPL William O. Wilson, Company C, 9th Cavalry
SGT William McGryar, Company K, 10th Cavalry

Spanish – American War

PVT Dennis Bell, Troop H, 10th Cavalry
PVT Fitz Lee, Troop M, 10th Cavalry
PVT William H. Thompkins, Troop G, 10th Cavalry
PVT George Wanton, Troop M, 10th Cavalry
SGM Edward L. Baker, 10th Cavalry

World War I
CPL Freddie Stowers, CO C, 371st Infantry

In 1988, the Secretary of the Army directed that the Army conduct research to determine whether there had been any barriers to Black soldiers in the Medal of Honor recognition process. The Army conducted extensive research during 1988 and 1989 at the National Archives and determined that Stowers was recommended for the Medal of Honor but, for reasons unknown, his recommendation was never processed. The Chief of Staff of the Army, Secretary of the Army, Chairman of the Joint Chiefs of Staff, and Secretary of Defense thoroughly reviewed the case file and recommended to the President that the Medal of Honor be awarded to Stowers.

World War II

No Black American military person received either the Army or Navy Medal of Honor for this war. Defense Secretary Frank Carluci initiated a review of the records of decorated Black servicemen during this war in an effort to determine if racial discrimination was a factor of denial so far as Black Americans receiving this medal. President William Clinton, in a ceremony on January 13, 1997, presented Vernon Baker, the only living recipient the Medal of Honor along with six other soldiers posthumously, thereby finally recognizing a total of seven Black servicemen for their heroic efforts in World War II.

- 1st Lieutenant Vernon J. Baker
- Staff Sergeant Edward A. Carter, Jr.
- 1st Lieutenant John R. Fox
- Private First Class Willy F. James, Jr.
- Staff Sergeant Ruben Rivers
- Captain Charles L. Thomas Private George Watson

Korean War

PFC William Thompson. CO M, 24th Infantry.

SGT Cornelius Charlton. Company C, 24th Infantry

Vietnam War

PFC Milton L. Olive III, Co B, 503rd Infantry
PFC James A. Anderson, Jr. Co F. 2nd Bn, 3rd Marine Division
SFC Webster Anderson, Btry A, 2nd Bn, 320th Artillery
SFC Eugene Ashley, Jr. Co C, 5th Special Forces Group
PFC Oscar P. Austin. Co E. 2nd Bn, 1st Marine Division
SFC Willian M. Bryant. Co A, 5th Special Forces Group
SGT Rodney M. Davis Co B, 1st Bn, 1st Marine Div
FC Robert H. Jenkins. Co C, 3rd Recon Bn, 3rd Marine Division
SP6 Lawrence Joel. HHC, 1st Bn, 503rd Infantry
PFC Ralph H. Johnson. Co A, 1st Recon Bn, 1st Marine Division
SP5 Dwight Hal Johnson, Co B, 1st Recon Bn, 1st Marine Division
PFC Garfield M. Langhorn. Troop C, 7th Sqd, 17th Cavalry
PSG Matthew Leonard, Co B, 1st Bn, 16th Infantry
SGT Donald Russell Long. Troop C, 1st Qd, 4th Cavalry
CPT Riley Leroy Pitts. Co C, 2nd Bn, 27th Infantry
LTC Charles C. Rogers. 1st Bn, 5th Artillery
1Lt Ruppert L. Sargent. HHC, 3rd Bn, 60th Infantry
SP5 Clarence E. Sasser, HHC, 3rd Bn, 60th Infantry
SSG Clifford C. Sims. Co D, 2nd Bn, 501st Infantry
1Lt John E. Warren, Jr. Co C, 2nd Bn, 22nd Infantry

United States Navy

Landsman Aaron Anderson, USS Wyandank, March 1865
Contraband Robert Blake, USS Marblehead, December 1863
Landsman William Brown, USS Brooklyn, August 1864
Landsman John Lawson, USS Hartford, August 1864
Engineer's Cook James Miffin, USS Brooklyn, August 1864
Seaman Joachim Pease, USS Kearsarge, June 1864
Seaman John Davis, USS Trenton, February 1881
Ship's Cook Daniel Atkins. USS Cushing, February 1898
Seaman Joseph B. Noil, USS Powhatan, December 1872
Fireman First lass Robert Penn. USS Iowa, July 1898
Seaman Alphonse Girandy, USS Tetrel, March 1901
Seaman John Johnson, USS Kansas, April 1872
Cooper William Johnson, USS Adams, November 1879
Seaman John Smith, US Shenandoah, September 1880
Seaman Robert Sweeney, USS Kearsarge, October 1881
Seaman Robert Sweeney, USS Jamestown, December 1883 (2nd Award).

The African-American Medal of Honor memorial was dedicated in Wilmington, Delaware on November 1, 1991. The memorial, in its own way, is a story of the sort of patriotic persistence demonstrated by Blacks in the service of this country. Wilson K. Smith Jr., a Black veteran who won the Bronze Star as an Army paratrooper in Vietnam, conceived and relentlessly pursued the memorial. The memorial is eight-sided, representing America's wars with a plaque on each side listing the names of Blacks who received the Medal of honor for that particular war. The base, with names and other information about the recipients, is completed but only two of the eight statues that will eventually stand on top have been finished. The completed project cost is estimated to be $135,000.00. When finished, the memorial will be permanently displayed in The Capitol Rotunda in Washington. If you would like to contribute or sponsor a fund raising event to help pay for completion of this memorial, contract Peoples Settlement Inc., Wilmington, Delaware, phone 1-302-658-4133.

This memorial to Black Medal of Honor recipients is only display at the Fort Leavenworth Post Exchange as part of the Buffalo Soldier Monument dedication program.

All of America's ethnic groups have contributed significantly to the freedoms this country enjoys. However, some groups have not received their just reward or recognition and, hopefully, each of you will do your part to correct this in the future. The Buffalo Soldier Monument Committee and its sponsors strongly urge Congress and the military to continue the review of all service records of Blacks who fought in World Wars I and II to identify other Blacks who were deserving but not overlooked. Write or call your Congressmen and Senators.

Fiddler's Green

When a cavalryman dies, he begins a long march to his ultimate destination. About half-way along the road he enters a broad meadow dotted with trees and crossed by many streams, known as "Fiddler's Green." As he crosses the green he finds an old canteen, a single spur, and a carbine sling. Traveling along he comes upon a field camp where he finds all the troopers who have gone before him, with the campfires, tents, and picket lines neatly laid out.

All other branches of service must continue to march without pause. The cavalrymen, though, are authorized to dismount, unsaddle and stay in the Fiddler's Green, their canteens ever full, the grass always green, and enjoy the companionship and reminiscences of old friends.

Author's Biography

Harold S. Cole, was born in North Pelham, New York. In 1942, he graduated from high school in New Rochelle, New York. After serving in the U.S. Army during World War II, he studied Horology at the Manhattan Training Center in New York and graduated in 1947.

In 1941, World War II was a reality. Like millions of young men, in 1942 he enlisted in the U.S. Army. His unit was Troop F, 9th U.S. Cavalry Regiment—a segregated unit with White Officers. The 9th Cavalry Regiment was organic to the 2nd Cavalry Division. In a short time, he became a platoon sergeant. Trooper Cole passed a test to become a Warrant Officer, was selected to attend Warrant Officer School, but his attendance did not happen. In early 1944, the 2nd Cavalry Division was transferred to the European Theater of Operations. After arrival in North Africa, the 2nd Cavalry Division was transferred to Sicily, and from there to Naples and Anzio, Italy. He boarded a ship for the invasion of Southern France and after that military operations, the war in Europe was almost over. He was then sent to Carentan, France, where he performed garrison duties until it was time to return to the United States.

In 1953, Trooper Cole enlisted in the U.S. Air Force and was sent to Aircraft School. Like the cavalry, serving in the Air Force was an excellent experience. He served in the Air Transport Command, Air Defense Command, Radar Early Warning System Command, and the Air Rescue Command. In 1956 – 1957, he participated in the construction of the North American Radar Early Warning System. In 1966, he participated in the recovery of aircraft from Vietnam.

Trooper Cole, retired from the military after twenty years of service. He served in seventeen countries and held a Top Secret security clearance. His awards from U.S. Army: U.S. Army Good Conduct Medal, American Campaign Medal, European-African-Middle Eastern Campaign Medal with two Bronze Stars, WWII Victory Medal with Germany Clasp, Belgian Fourragere, WWII Honorable Service Lapel Button. U.S. Air Force—Air Force Longevity Service Award Ribbon with three Oak Leaf Clusters, Air Force Outstanding Unit Award with two Oak Leaf Clusters, National Defense Service Medal with one Bronze Star, Air Force Good Conduct Medal with three loops.

After retirement from the U.S. Air Force, Trooper Cole worked at Grumman Aircraft, Fairchild Aircraft, and Lockheed Aircraft. He also retired from Lockheed Aircraft.

Trooper Harold S. Cole, had the pleasure of meeting Edwina M. Roberts, in 1984. They were married and Harold has four very fine stepchildren.

Army of the United States

Honorable Discharge

This is to certify that

HAROLD S COLE   12 181 828   TECHNICIAN FOURTH GRADE
TROOP F   -   9TH CAVALRY

Army of the United States

is hereby Honorably Discharged from the military service of the United States of America.

This certificate is awarded as a testimonial of Honest and Faithful Service to this country.

Given at SEPARATION CENTER
FORT DIX   NEW JERSEY

Date 9 JANUARY 1946

J. H. GUNTER
MAJOR INFANTRY

Honorable Discharge

from the Armed Forces of the United States of America

This is to certify that

STAFF SERGEANT HAROLD S. COLE AF 12 181 828 REGULAR AIR FORCE

was Honorably Discharged from the

United States Air Force

on the 21ST *day of* OCTOBER 1963. *This certificate is awarded as a testimonial of Honest and Faithful Service.*

DOROTHY J. SCOTT
2ND LT., USAF

256 AF PREVIOUS EDITIONS OF THIS FORM MAY BE USED.
THIS IS AN IMPORTANT RECORD — SAFEGUARD IT!

Harold S. Cole
3840 Dusty Glen Ct.
North Las Vegas NV 89032-3172

Dear Harold S. Cole:

America has always called upon the Army to do the hard jobs, and her Army has always answered the call. We are a free Nation, a Nation looked upon by billions of people around the world as the standard-bearer of freedoms -- freedoms guaranteed by our great Soldiers and by the Veterans who have served our great country in the active Army, the Army National Guard and the Army Reserve.

We are an Army of tradition. Our greatest tradition is service to our country, and in that service we ask our Soldiers to carry out tough missions all around the world. Today, answering the Nation's call may take them into harm's way in our Global War on Terrorism, a war that touches us all and one that we must win.

As today's Soldiers serve, the traditions and legacy passed on by Army Veterans in previous conflicts sustain them so they can do what must be done to protect our freedoms. Our Army would not be the best in the world without the service of its Veterans who continue to provide support and exhibit patriotism. You make a difference to today's Army and to our country.

As a tribute to your service in the U.S. Army, The American Legion Department of Nevada has recognized you for this special Freedom Team Salute Commendation. It is our honor to bestow it and the enclosed Army lapel pin to you. The Commendation symbolizes the partnership between our Army, her Soldiers, their Families, and Veterans -- a partnership as old as the Nation itself. It is a partnership America has always been able to depend on. As our partner, we hope that you will display this Commendation and wear this pin with pride, as a statement of our shared commitment to support America's Soldiers.

We thank you for the honor of your past service with our Army and for your continued support of our Soldiers.

Sincerely,

George W. Casey, Jr.
General, United States Army
Chief of Staff

Pete Geren
Secretary of the Army

Trooper Harold S. Cole's military medals and awards

THE WHITE HOUSE
WASHINGTON

April 20, 1995

Greetings to the veterans of the Ninth & Tenth (Horse) Cavalry Association as you gather for your reunion.

Your units served the United States with honor and distinction during a crucial period in our history. We owe our liberties to the sacrifices of people who, like you, were willing to risk their lives for freedom. I know you join me in honoring your fallen comrades.

Each of you embodies the pride, professionalism, and accomplishment that make the United States Army one of the finest fighting forces the world has ever known. I salute you for your distinguished record of service, and I hope that you will enjoy your time together as you reflect on the bonds you share.

Best wishes for a memorable reunion.

Bill Clinton

MAXINE WATERS

MEMBER OF CONGRESS

35TH DISTRICT CALIFORNIA

BANKING AND FINANCIAL SERVICES COMMITTEE

JUDICIARY COMMITTEE

July 31, 2002

Trooper Harold S. Cole
Buffalo Soldiers
Ninth & Tenth (Horse) Cavalry Association

Dear Trooper Cole:

Congratulations to the Buffalo Soldiers, Ninth & Tenth Cavalry Association, Greater Los Angeles Area Chapter, on its 136th Anniversary Reunion.

This event provides the community an opportunity to recognize the contributions and achievements of the Ninth & Tenth U.S. Cavalry Regiments in the service of our nation.

My best wishes for a truly inspirational event.

Sincerely,

Maxine Waters

MAXINE WATERS
MEMBER of CONGRESS

DEPARTMENT OF THE ARMY
HEADQUARTERS, 101ST AIRBORNE DIVISION (AIR ASSAULT) AND FORT CAMPBELL
FORT CAMPBELL, KENTUCKY 42223-5000

February 26, 2001

REPLY TO
ATTENTION OF:

Trooper Harold Cole
3700 Lennox Ave #C-12
Van Nuys, CA 91405

Dear Trooper Cole:

    Sir, I felted compelled to write a brief note thanking you again for your visit to the Screaming Eagles and the Fort Campbell community. We all were truly blessed to have the opportunity to sit and talk with you and hear your presentations over the four days of your visit. I especially enjoyed the sincere and deep conversations we had riding from event to event and sharing over meals. I will never forget you and I want to again thank you for lifting my spirits and encouraging me.

    I have had several parents whose kids heard you speak at their school say to me, what an impact the group had on their child. If there is ever anything in which we can do for you, please let us know. If it is in our power to accomplish it, rest assured that we will get it done.

    We have an old saying in the Division, **"once a screaming eagle, always a screaming eagle"**. Since the Commanding General made you screaming eagles with the caps, coins and pocket knives, consider yourselves screaming eagles and you do not need an invitation to visit again, you are always welcomed!

Take care and May GOD's richest blessing continue to fall upon you and your family.

                            Sincerely,

                            Keith C. Blowe
                            Lieutenant Colonel, U.S. Army
                            Division Equal Opportunity Program
                            Manager

Greater Los Angeles Area Chapter
of
9th & 10th Cavalry Association

Buffalo Soldiers

Bottom row-left to right:
Trooper Jean Lewis, Trooper Orlanders Elliott,
Trooper Ernest Collier, Trooper Afred Evens,
Trooper Royal Carter, Trooper Curtis James,
Trooper Robert McDaniel, Trooper Harold Cole
Middle row-left to right:
Trooper Oliver Muldrew, Jr., Trooper Johnnie Mitchell,
Trooper Andrew Isaacs, Trooper George Poston,
Trooper Eugene Lewis, Trooper C.R.Troy Walker,
Trooper Walter Brady, Trooper Braxton Berkley,
Trooper Chris Vasgas, Trooper Fred Jones
Top row-left to right:
Trooper Ron Jones, Trooper Bruce Dennis, Trooper Charles Allen,
Trooper Virgil Griffin, Trooper Lennister Williams,
Trooper Fred Catha, Trooper Waldo Henderson

Photographer & Chapter member -Trooper Edwina Roberts-Cole

We Salute the National Council of Negro Women

"THANK-YOU-AMERICA CERTIFICATE"
1944 - 1945

Letter from the Ambassador of France,
François Bujon de l'Estang,
to the Heads of Veterans Associations in the U.S.

- Thank-You-America
- American Role in WWII
- Ambassador's letter
- Apply

Mr President
Veterans Association

Dear Mr. ...,

It is my pleasure to inform you that the French authorities have decided to issue a certificate to recognize the participation of all American and allied soldiers who took part in the Normandy landing and contributed to the liberation of France.

In agreement with the Acting Secretary of Veterans Affairs, The Honorable Hershel W. Gober, I come to you to request your association's assistance in identifying all eligible veterans. The certificate is meant to express the gratitude of the French people to the soldiers who participated in the Normandy landing and liberation of France, on French territory and in French territorial waters and airspace, between June 6, 1944 and May 8, 1945. The certificate will not be issued posthumously.

The ten Consuls General of France in the United States will issue the certificates on behalf of the French authorities. They will do so in coordination with State Veterans Affairs Offices, Veterans Service Organizations' national and states representatives and, hopefully, with the assistance of your association and other Veterans organizations.

Indeed, owing to the huge number of qualifying veterans, it will be necessary for us to work through this project with Veterans associations and organizations and their members on a volunteer basis, in each of the fifty states, in order to identify the qualifying veterans, review and certify the applications, prepare the certificates and organize the ceremonies to present them to the veterans.

To orient our work, it would be extremely useful that your representatives in each state contact the Consuls General of France--whose list is attached to this letter--so that they can start working with them state by state. You will find enclosed the application form and instructions on how to apply. More information is also available on the Embassy's web site: http://www.ambafrance-us.org. You will also find herewith a short presentation of the project for the purpose of information of your members.

Thanking you in advance.

Yours sincerely,

François Bujon de l'Estang

Search   About Us   Faq
Site Map   Home   Contact Us

Embassy of France, December 4, 2000

Liberté • Égalité • Fraternité
RÉPUBLIQUE FRANÇAISE

CONSULAT GENERAL DE FRANCE A LOS ANGELES

Los Angeles, July 17th 2001

Dear Sir:

I have received the copy of your certificate of honorable discharge for which I thank you.

I am very pleased to enclose the Certificate of Appreciation issued by the French Secretary for Veteran Affairs and the President of the Basse Normandy Region to all American veterans who took part in the Normandy landing and the following combats for the liberation of France.

I would like to take this opportunity to thank you myself and on behalf of the French citizens for your sacrifices during the war.

Wishing you the best, I remain,

Very sincerely,

Michel CHARBONNIER
Deputy Consul General

Serve, Preserve, Honor

October 5, 2007

Harold S. Cole
3840 Dusty Glen Court
North Las Vegas, NV 89032

Dear Mr. Cole,

    The Library of Congress Veterans History Project has recently received the interview you did for our project collected by BG Don Scott, USA (ret). As the director of the project, I want to thank you for taking the time to record your wartime remembrances. I am certain they will be a valued part of our collection and will serve as an inspiration for generations.

    Again, thank you for your participation in this project and more importantly thank you for your service to our nation.

Yours truly,

Bob Patrick
Director

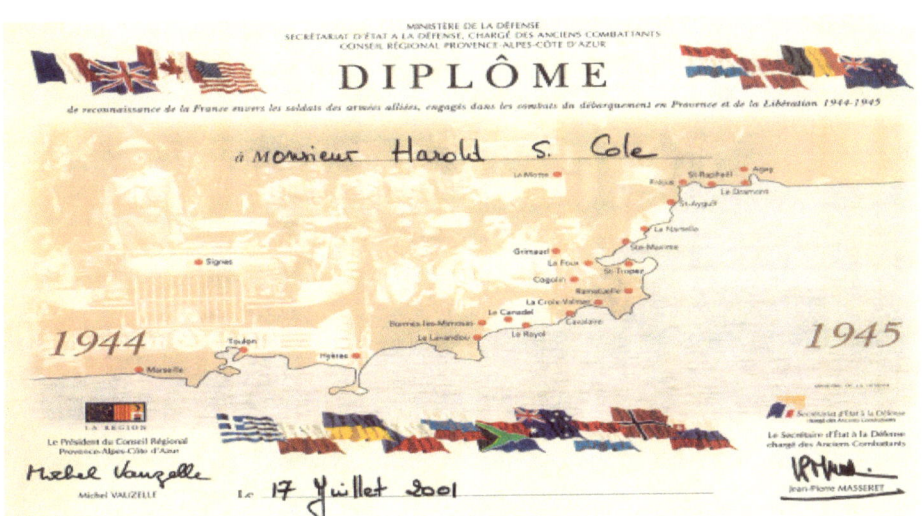

A Psalm of David

The Lord is my shepherd;
I shall not want.

He makes me to lie down in green pastures;
He leads me beside the still waters.

He restores my soul;
He leads me in the paths of righteousness
For His name's sake.

Yea, though I walk through the valley of the shadow of death,
I will fear no evil;
For You are with me;
Your rod and Your staff, they comfort me.

You prepare a table before me in the presence of my enemies;
You anoint my head with oil;
My cup runs over.

Surely goodness and mercy shall follow me
All the days of my life;
And I will dwell[a] in the house of the Lord
Forever.

Psalm 23

Got an idea for a book? Contact Curry Brothers Books, LLC. We are not satisfied until your publishing dreams come true. We specialize in all genres of books, especially religion, leadership, family history, poetry, and children's literature. There is an African Proverb that confirms, "When an elder dies, a library closes." Be careful who tells your family history. Are their values your family's values? Our staff will navigate you through the entire publishing process, and take pride in going the extra mile in meeting your publishing goals.

Improving the world one book at a time!

Curry Brothers Books, LLC
PO Box 247
Haymarket, VA 20168
(719) 466-7518 & (615) 347-9124
Visit us at www.currybrothersbooks.com

www.ingramcontent.com/pod-product-compliance
Lightning Source LLC
Chambersburg PA
CBHW042058290426
44113CB00001B/11